ELECTRICIANS
Coloring Book

Jasmine Taylor

ISBN 978-0-359-86490-4

9 780359 864904

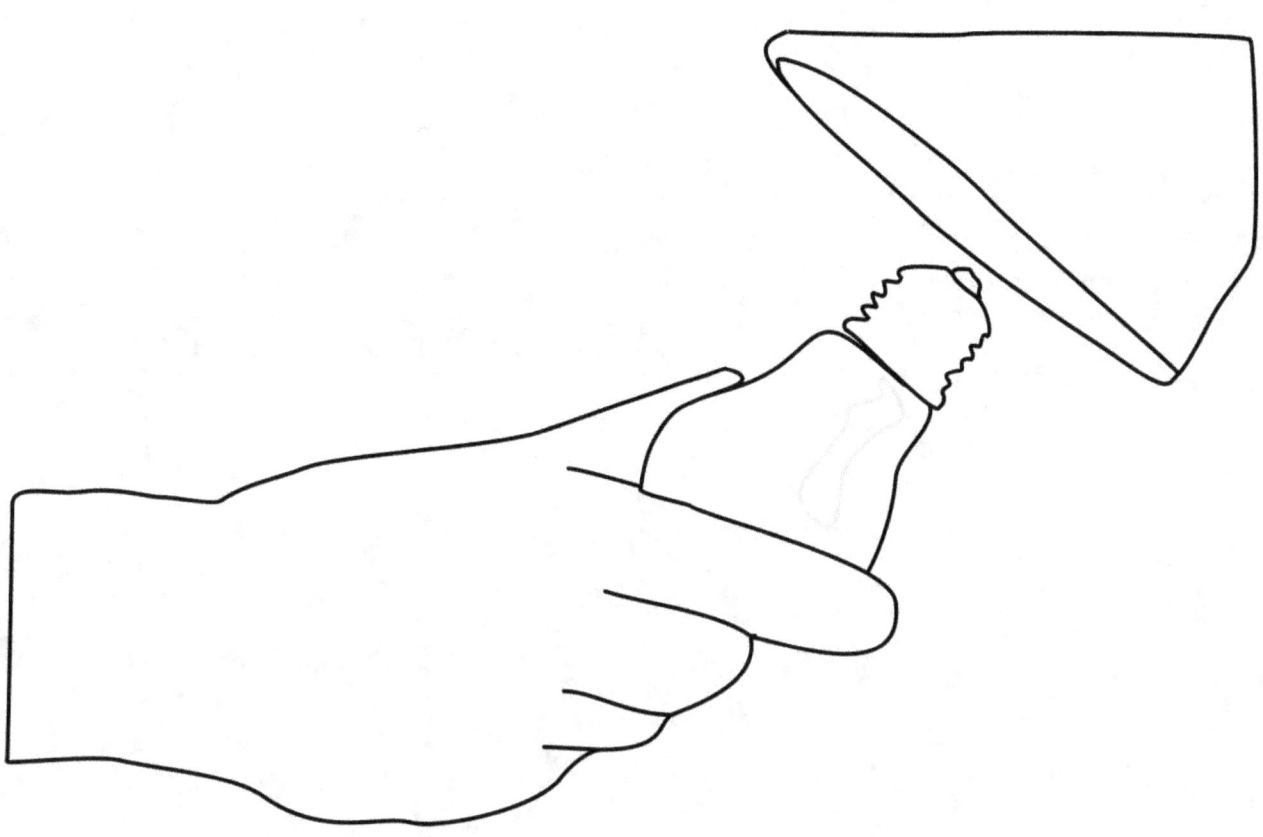

ARTIST

_____
NAME

_____
DATE

SERVICE

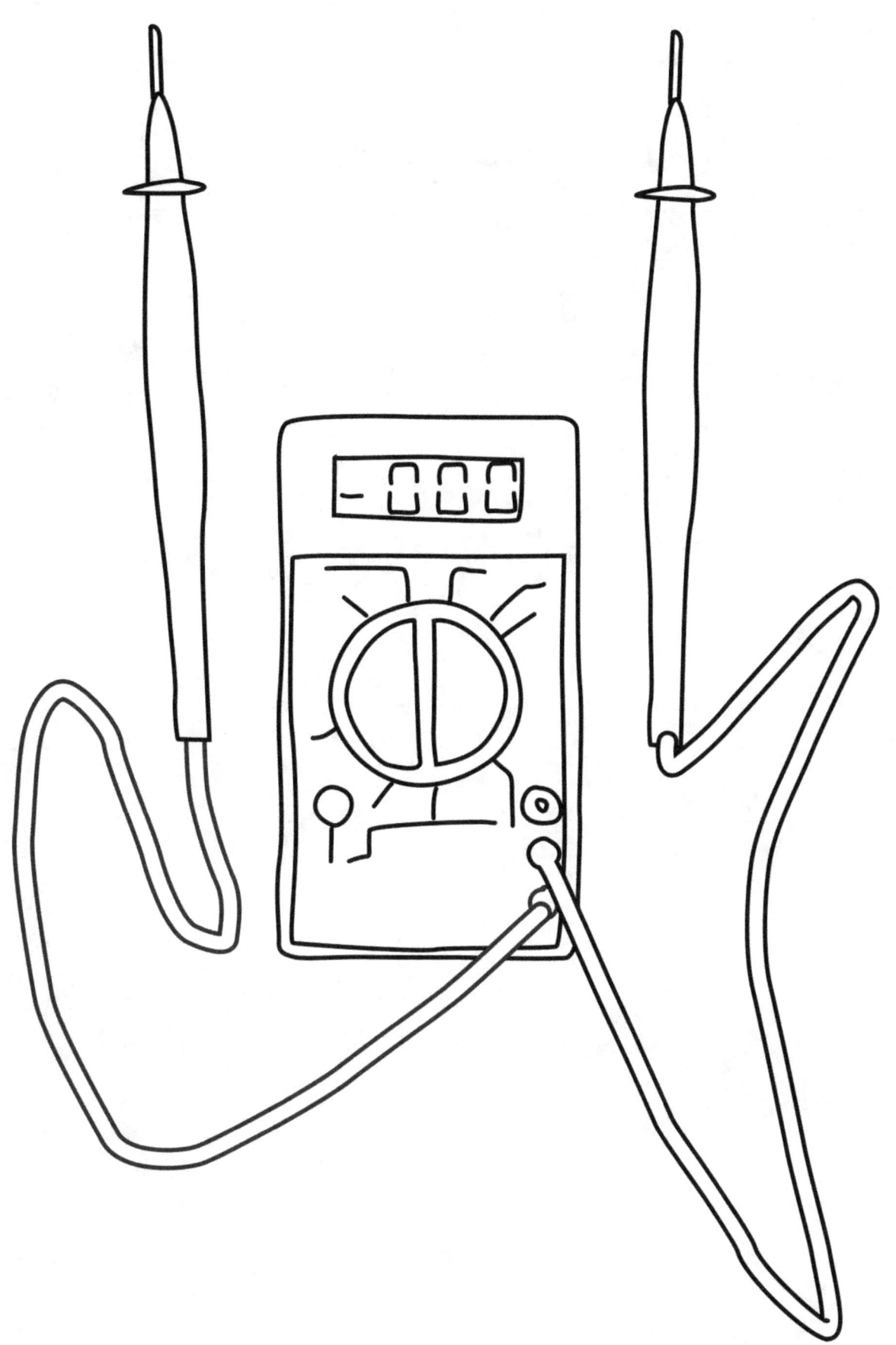

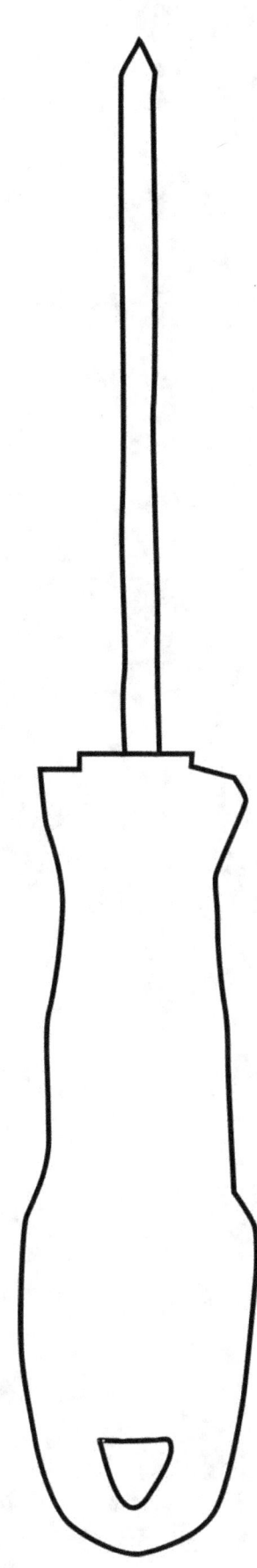

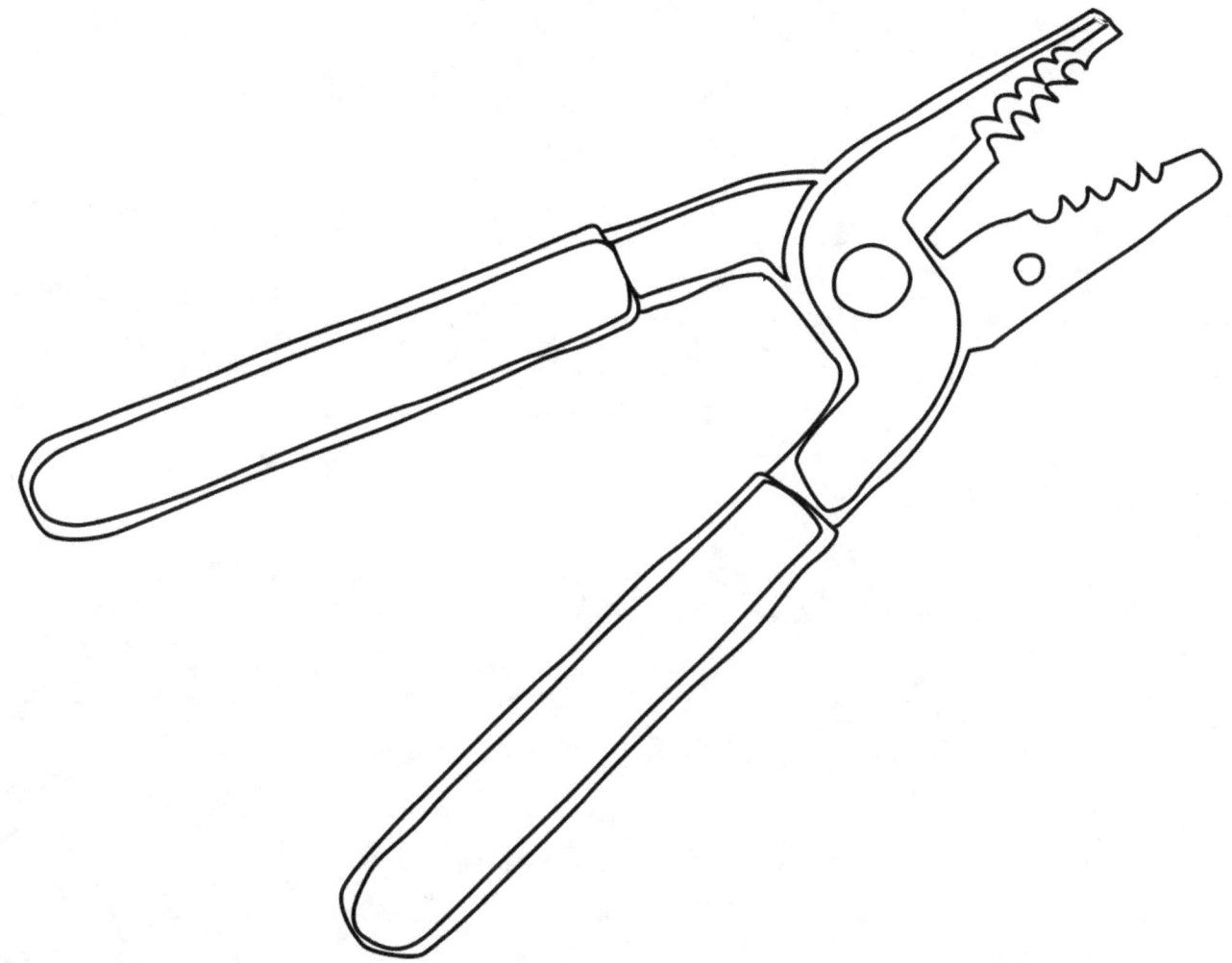

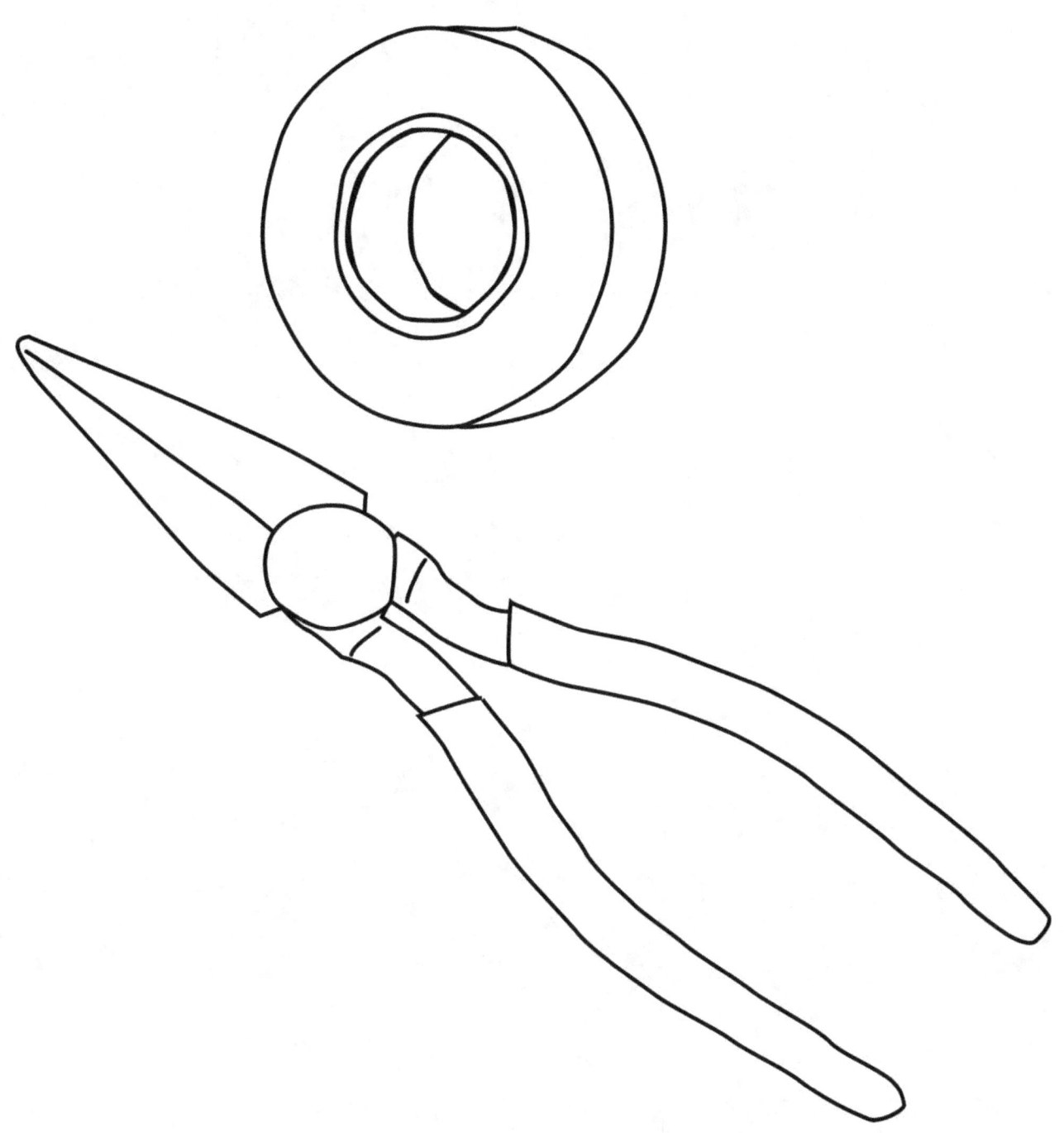

ARTIST

_____

NAME

_____

DATE

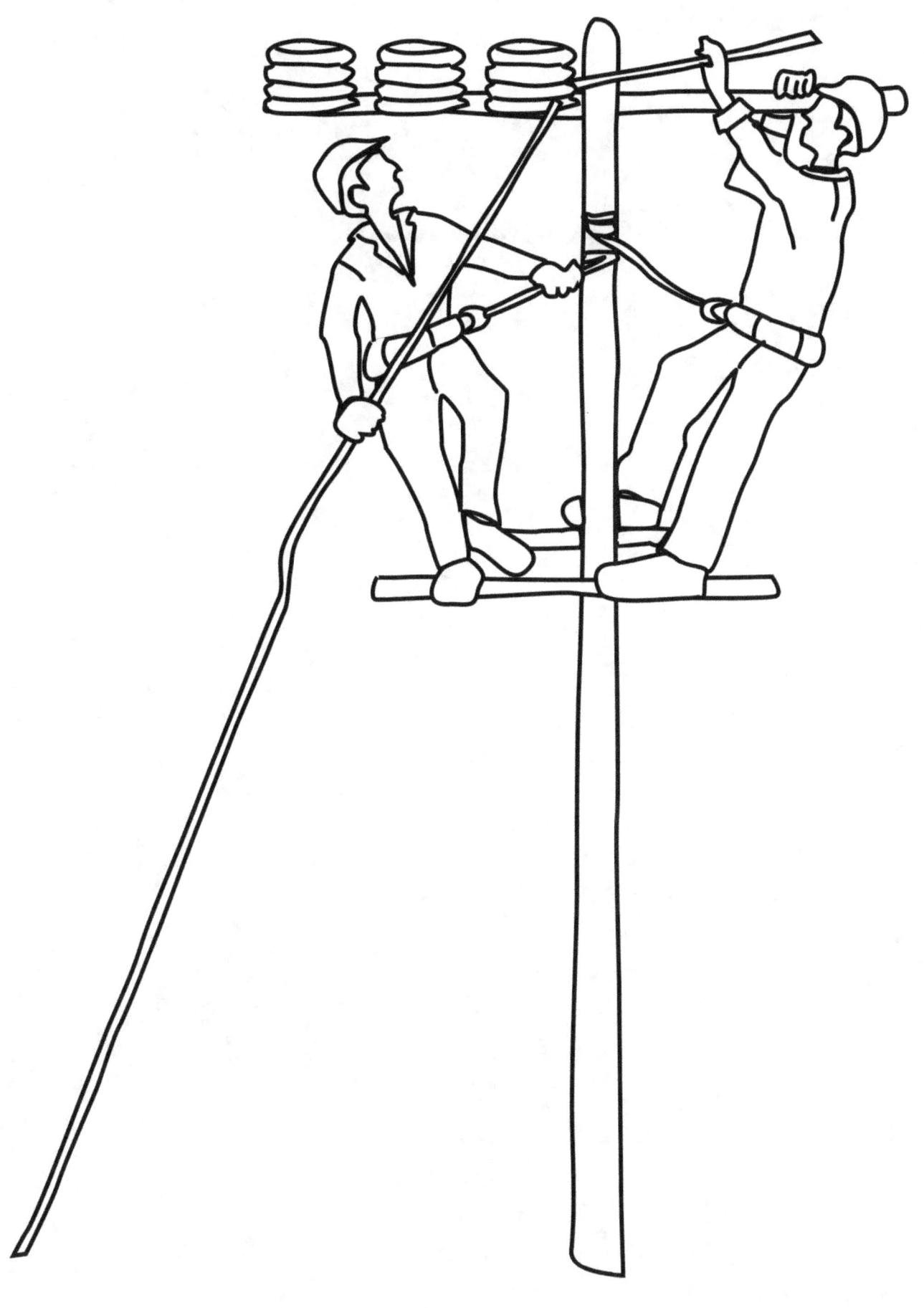

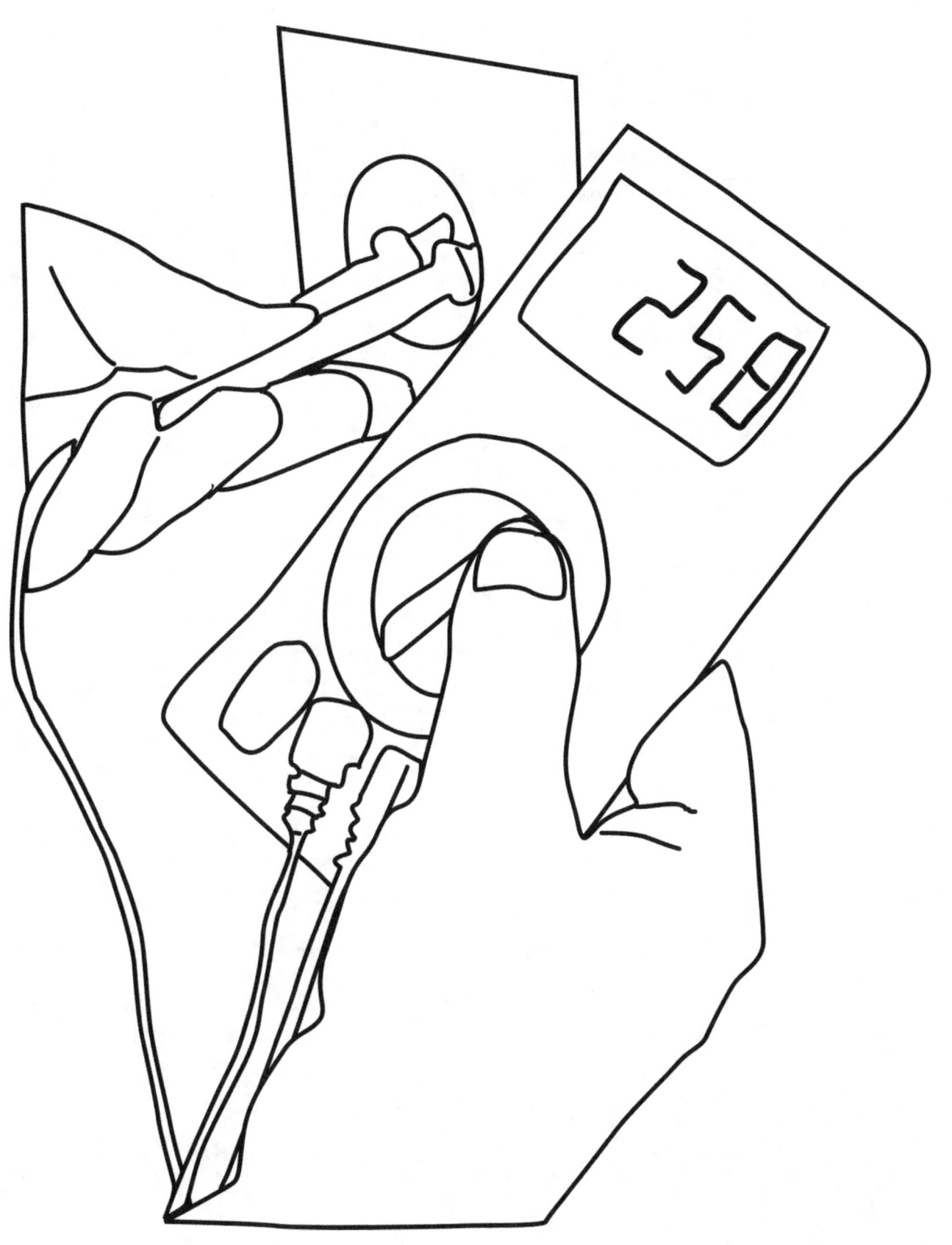

ARTIST

_____
NAME

_____
DATE

ARTIST

_____

NAME

_____

DATE

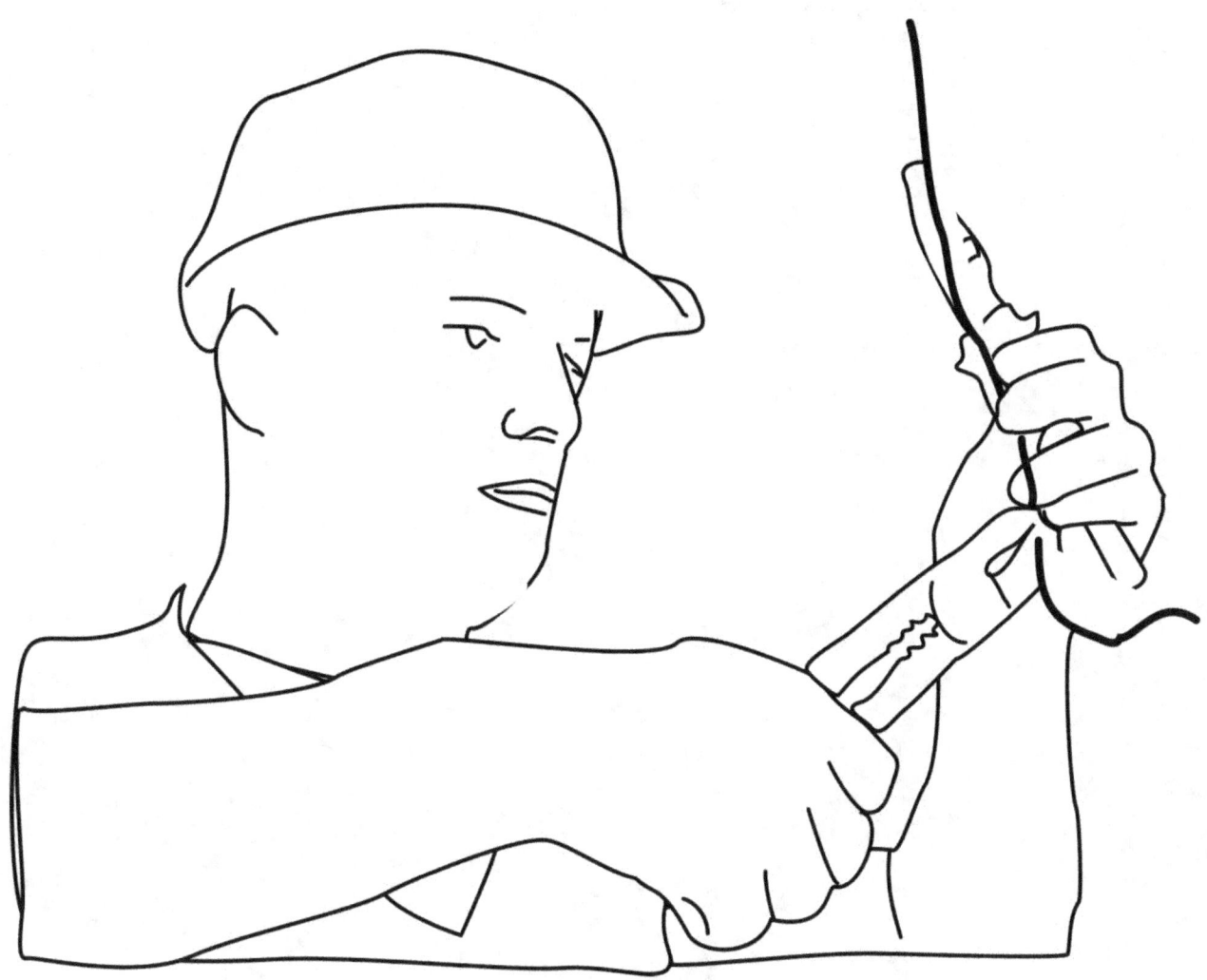

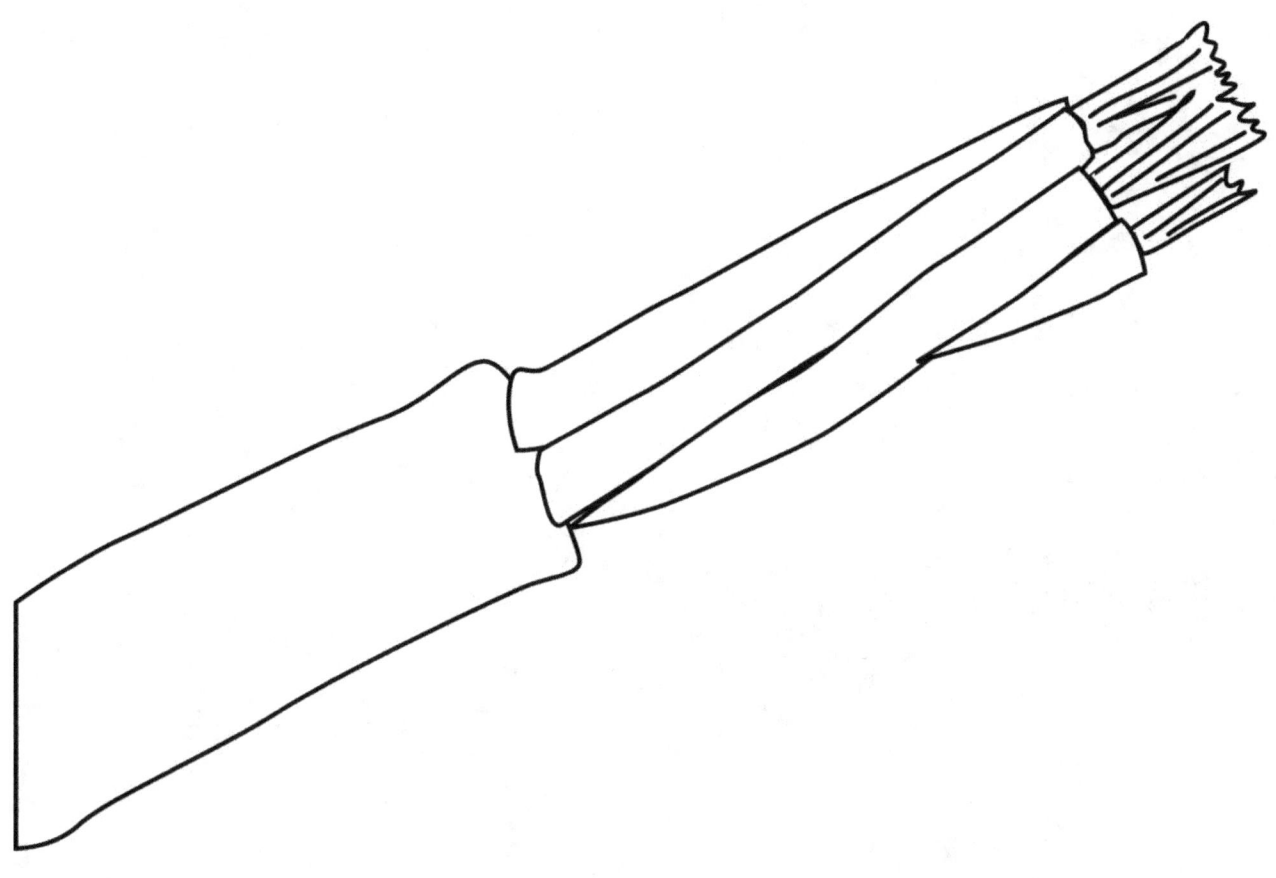

ARTIST

_____
NAME

_____
DATE

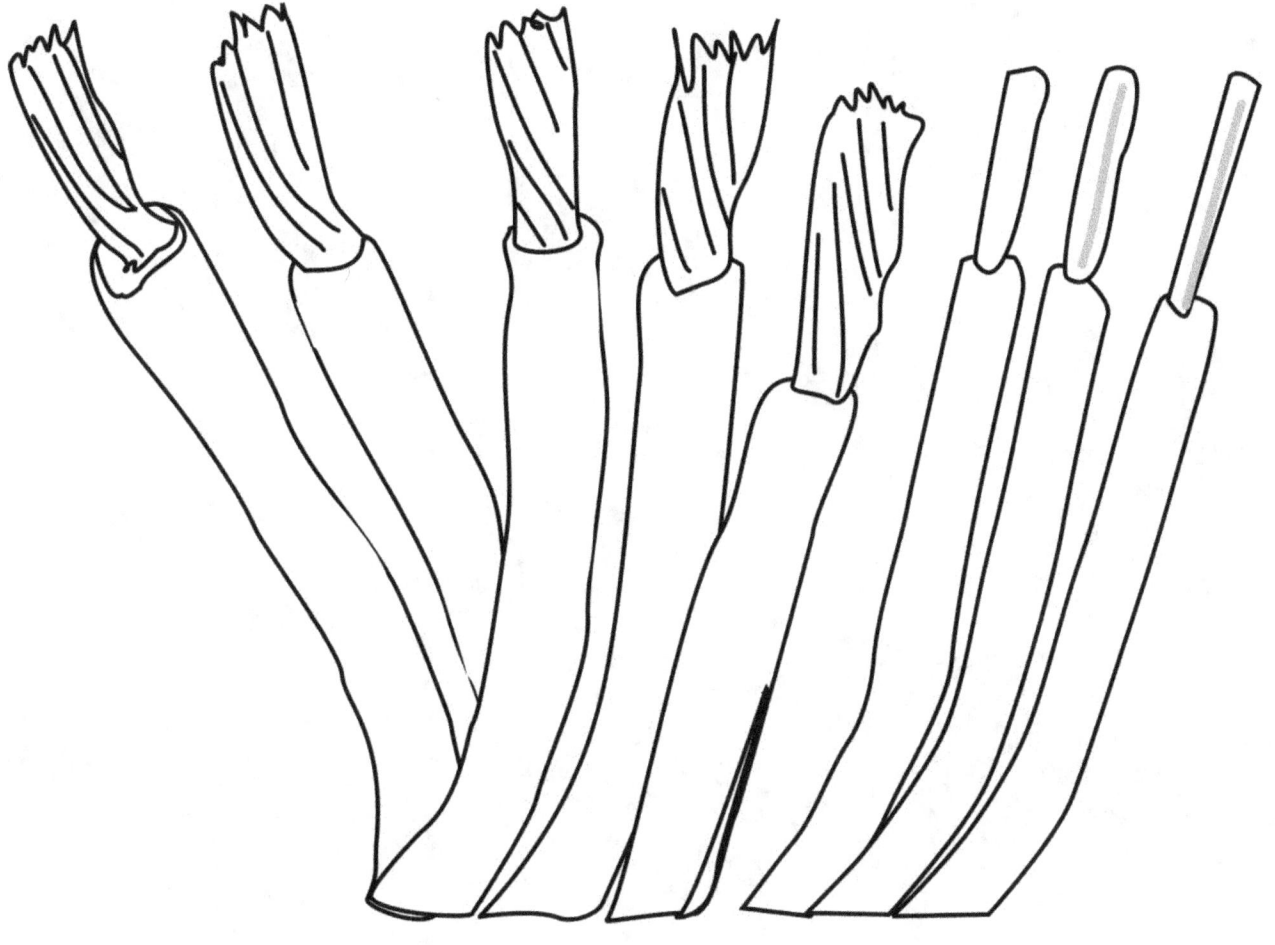

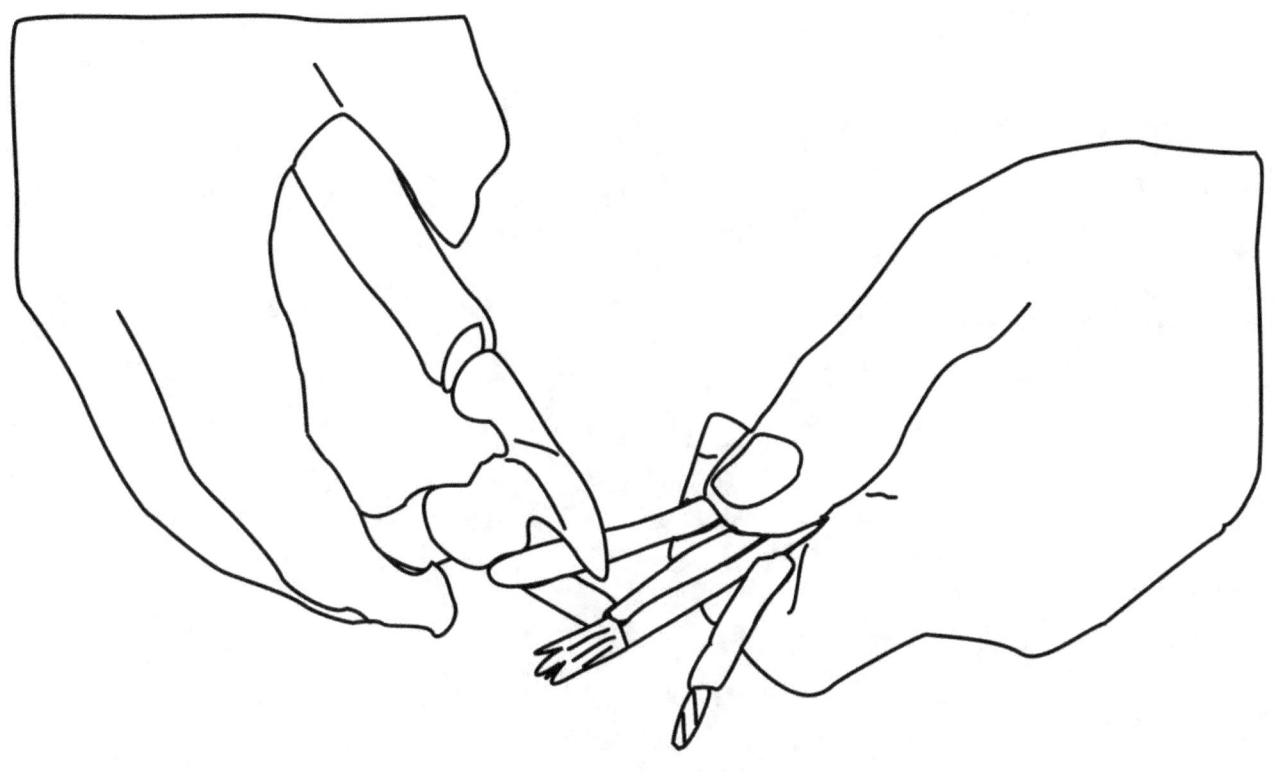

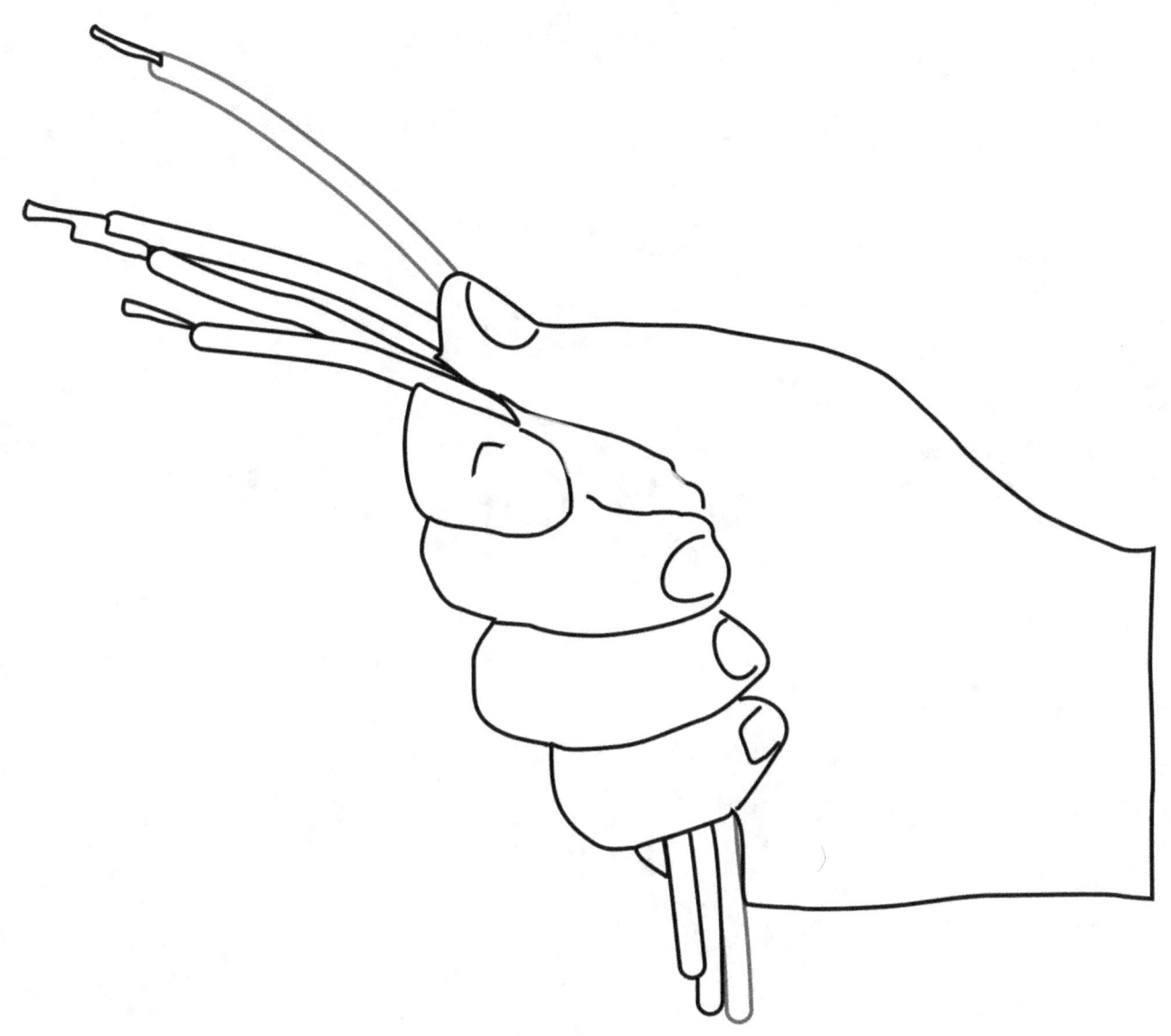

ARTIST

_____
NAME

_____
DATE

Artist Notes

Other Products

Please visit Jasmines author page for other coloring book titles.

www.amazon.com/author/jasminetaylor (https://goo.gl/AYQdxw)

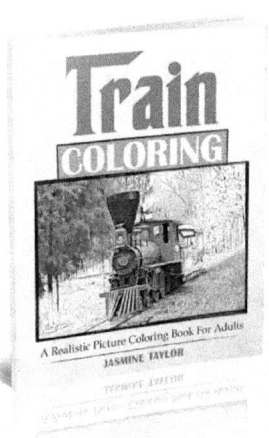

About the Author

I'm passionate about art, coloring and creativity. My coloring books cover a variety of niches with varying styles and levels of difficulty from beginner to advance for all to enjoy. I'm really thrilled to offer my coloring books to you in the hope of inspiring your creative journey.

If you enjoyed this coloring book, please help others find and benefit as well by leaving a positive review on Lulu or one of its distribution partners. Thank you very much and happy coloring!

Jasmine Taylor